64 Secrets Still Ahead of Us

Ways in Which an Earlier Science and Technology was Superior to Today's

By
Jonathan Gray

TEACH Services, Inc.
New York

2006   07   08   09   10   11   12   ·   5   4   3   2   1

Copyright © 2002 Jonathan Gray
Copyright © 2006 TEACH Services, Inc.
ISBN-13: 978-1-57258-418-1
ISBN 1-57258-418-1
Library of Congress Control Number: 2006920053

Published by

TEACH Services, Inc.
www.TEACHServices.com

CONTENTS

3

Town planning

Mechanical – electronic

Everyday items

Clothing, textiles

Art, sculpture

Health, medical

Electricity

Surveying

Flight

Intriguing secrets

I think you'll agree, there are some interesting items on that list. In the next few minutes, we shall look at each of these in turn. But first I would like to share with you an event from America's early history...

INTRODUCTION

The young man from South Dakota attracted no particular attention when he arrived in Ouray, nor did he wish to. Ouray in the days of the silver boom was used to having unattached young men arrive to take jobs in the mines, to look the country over, or to exist without any seeming means of support. No one asked any questions about another's business; if they had done so it might not have been safe, and it certainly would not have been considered good manners.

So this young man from the the north cleaned out an abandoned cabin, settled down, and minded his own business. He went into the town for his meals morning and evening, and what he did during the day no one knew or cared.

If he had been watched he would have been seen climbing over the almost perpendicular mountains which enclose Ouray on three sides, giving special attention to the abandoned silver mines, where the strain of ore had run out. In the evening he would come in with a sack full of rocks slung over his shoulder—a common enough sight, and not one to cause attention.

Presently he received a number of assay reports from Denver, and in consequence he made some contacts with the owners of several of these worked-out properties. Still Ouray did not know what he was doing. After all, it was a silver-mining town, and those properties, as everyone knew, were worthless for mining. It is probable that the owners who sold out to the youth chuckled to themselves at the time because he had paid good money even for a mere option on such worthless propositions.

But they did not chuckle, nor did the townspeople continue to disregard him, when, a few weeks later, he began mining operations for gold in those same abandoned properties. For that is just what he had found! He had made a rich—fabulously rich—strike. The very ore dump at the Camp Bird Mine assayed $3,000 to the ton. In today's terms, that would be worth many times more.

Thus did young David Ignatius Walsh, later United States Senator from Massachusetts, find wealth in a territory that had never been thought of as gold country, because he had studied a little metallurgy and had observed that the same sort of rock that was being discarded had gold in it in South Dakota.

The experienced silver miners of the area had been walking over a king's ransom, had been cursing it, throwing it out on the ore dumps. The gold had been there all the time, but no one had taken time to search for it and take advantage of it.

ARE WE MISSING SOMETHING FROM THE PAST?

You and I can learn from this experience – if we will just stop and think on the matter.

Our modern world is flattering itself that never before in the history of man has there been such a brainy lot of scientists as are with us now.

In this 21st century we are so educated, so cultured, so advanced, that it may be tempting to think that our ancestors were by comparison bumbling cave men.

Could this be a simple case of egotism, based on ignorance?

Yes, we do have wonderful technology today. And it is exploding at a rate that is breathtaking. But is this because we are more intelligent than our ancestors...or could there be another reason?

I think most of us will realise, mankind is no more intelligent than he was a thousand years ago, but there is a factor in our favour: we have accumulated more technology. We have the accumulated knowledge of the past upon which we can draw and make improvements.

That's why we have so much. And now our accumulated knowledge is building exponentially. What an exciting time to be alive!

UNCLAIMED TREASURE

But, like the townsfolk of Ouray, many of us have been walking over our past, regarding it lightly, unaware of priceless treasures – lost, discarded and forgotten. Literally there is a king's ransom of information that could transform our life style in many other ways, if we would only search, study and take advantage of it.

I did not come upon this information by choice. A few years ago I happened to be researching for a history book, in which prehistory would have probably occupied just one brief chapter.

At first, when I stumbled across one "out of place" item, I simply dismissed it as irrelevant.

However, it was not just in one spot. As I investigated further, a few more of them popped up...in various places. Still, I continued to push them aside...because they could not be explained.

These "out of place" items continued to surface. Now they were emerging more persistently. Finally, it became clear. A pattern was developing.

You see, most archaeologists specialise in the country of their interest, and become proficient in Egyptian archaeology, North American, African, or whatever. And the odd "out of place" find is usually dismissed. It is only when you start taking in the BIG picture—the world-wide scene—when you notice the same "out of place" items in other parts of the world, that it becomes possible to see a pattern emerging.

I discovered that there was a connection between these "out of place" items—whether in South America, the Pacific or the Old World. There was an "agreement" between them. They spoke with the same voice...hinting at a prehistoric high civilisation, a global super culture, that had been lost, ignored, or forgotten.

Here was something that we had never been told about! This was the missing ingredient to history! Suddenly it struck me. That's when I resolved to take the matter more seriously. So I scrapped plans for the book I was writing...and began serious research for what would eventually become *Dead Men's Secrets*.

And now, a natural fruitage of this—the present survey.

Do you know that most information concerning ancient secrets has been lost? Do you know that most of it was deliberately destroyed! Do you know that over the centuries there has been widespread destruction of artifacts and printed records, in which the greatest depositaries of knowledge from the ancient world are lost forever? Fortunately, a small portion did survive. You have to search for it. But it's there! And some of it is quite mindblowing!

THREE CLASSES OF INFORMATION

There are three ways to track this information.

Firstly, through PHYSICAL ARTIFACTS. There are objects of metal sitting in museums, unquestionably made in the ancient world, that would have required very advanced technology to produce. Technology not to be repeated until our day. The weight of evidence grows daily – evidence that all the major secrets of technology were known, and forgotten, long ago. Evidence that early man did create a society that surpassed ours in all aspects of development.

Secondly, WRITTEN RECORDS. In addition to these artifacts, we have some priceless ancient books. These go into detail concerning a superior knowledge in the remote past...knowledge concerning the surface of the moon, the atomic structure of matter and so on. Some of this information is still ahead of our time.

Thirdly, ORAL TRADITIONS. There is a huge mass of oral traditions – a shared racial memory – concerning an original superior civilisation and its scientific secrets. I am reminded of the question that was posed by the eminent British scientist Frederick Soddy, winner of a Nobel prize in physics. He asked whether ancient civilisations might "not only have attained our present knowledge, but a power hitherto unmastered by us?" (Soddy, Interpretation of Radium, p. 243)

THEY KNEW MORE

Certainly, as evidence piles up of a forgotten prehistoric science, it is difficult to shake off the feeling that our ancestors knew a lot more than we do.

They possessed superior intelligence and technological skill—often to a degree that the modern mind finds staggering. Their attainments stare us in he face. They are astonishing; in some spheres, penetrating knowledge which our science has scarcely begun to nudge. Their secrets defy us.

Now let us get down to some specifics. That will be the purpose of this survey. I tracked down the following examples, in the hope that you will find them interesting.

We shall touch on <u>sixty-four</u> aspects of science and technology in which the ancients were superior to us.

TREASURES OF KNOWLEDGE
AHEAD OF OUR TIME

CARTOGRAPHY

1. Maps more accurate than ours

I would like to mention the <u>Zeno map</u>, drawn in 1380. It accurately outlines the coasts of Norway, Sweden, Denmark, Germany and Scotland, as well as the exact latitude and longitude of a certain number of islands.

Now, the chronometer, necessary to determine longitude, was not invented until 1765. Nevertheless, the Zeno map is most accurate. The topography of Greenland is shown free of glaciers as it was prior to the Ice Age. Unknown rivers and mountains shown on this Zeno map have since been located in probes of the French Polar Expedition of 1947–1949.

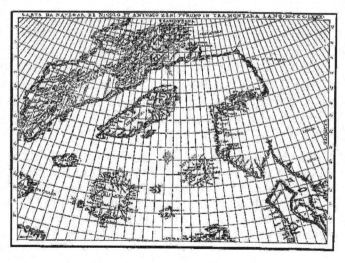

You ask, How can this be? Where did Zeno get his information?

The truth is, many of the Medieval and Renaissance mapmakers admitted they were copying from sources whose origins were unknown.

These maps are a scientific achievement far surpassing the abilities of the navigators and mapmakers of the Renaissance, Middle Ages, the Arab world, or any ancient geographers. They are the product of an unknown people antedating recognised history.

Then there's the Zauche map of 1737. It shows Antarctica. Remember, Antarctica's existence was not verified until 1819!

This map not only shows Antarctica, but shows it completely free of ice. Surprisingly, it is shown not as one continent but two islands separated by a strait from the Ross to the Weddell Seas (a fact which was not established until the Geophysical Year, 1968). Also shown are islands of the Mid-Atlantic Ridge, now known to lie on the bottom of the ocean.

You have probably heard by now of the Piri Reis chart of 1513. After its discovery, Captain Arlington H. Mallery, an American authority on cartography, asked the U.S. Hydrographic Office to examine it. The U.S. Navy, through Commander Larsen, subsequently issued this statement:

> The Hydrographic Office of the Navy has verified an ancient chart— it's called the Piri Reis map, that goes back more than 5,000 years. It's so accurate, only one thing could explain it—a worldwide survey. The Hydrographic Office couldn't believe it, either, at first. But they not only proved

the map genuine, it's been used to correct errors in some present-day maps.

If ever there were a treasure map, this is it. Just crammed with priceless gems. It tells the story of ancient coastlines, as well as the surprising exploits of our ancestors five thousand years ago.

Piri Reis stated that his copy was a composite from twenty ancient maps.

This ancient chart shows South America and Africa in correct relative longitude and latitude. Not only were the Caribbean, Spanish, African and South American coasts charted in correct positions relative to each other, but even isolated land areas, like Cape Verde Island, the Azores, the Canary Islands, as well as topographies of the interiors—mountain ranges, peaks, rivers, plateaus. All were accurately positioned by longitude and latitude.

This map shows the coastline of Queen Maud Land in Antarctica, with its islands and bays the same as they appear below the Antarctic ice sheet (as revealed by modern seismic echo soundings). Pictured in great detail are regions scarcely explored today, including a mountain range that remained undiscovered until 1952. Interestingly, the map shows two bays where the modem seismic map showed lands. However, when the experts were asked to check their measurements, they found that the ancient map was correct, after all.

It now becomes crystal clear. Either somebody mapped Antarctica before the ice cap covered the continent, or else the ice-covered continent was mapped with very sophisticated instruments.

A major error appeared to be Greenland, shown as three islands. But during the International

Geophysical Year it was proved that this correctly represented the state of affairs about 3000 B.C.

Every mountain range in northern Canada and Alaska was recorded on this ancient map—including some ranges which the U.S. Army Map Services did not have on their maps. But the U.S. Army has since found them!

There are distortions on Piri Reis' map. The ancient original source-maps were drawn using a circular grid based on spherical trigonometry, with the focal point situated in Egypt. The copiest Piri Reis (unfamiliar with circular projection) shifted and spliced the original grid to compensate for the curvature. Any modern spheroid projection on a flat surface would cause the same distortion.

The evidence indicates that what we have here is only part of an original world map.

So here is evidence of science in an early epoch, which is considered to have had none. Here are physical fragments of the amazing knowledge of a super culture long vanished.

2. Complex, 3-dimensional map made with unknown technology

A find made by Russian scientists runs counter to traditional notions of human history. It is an ancient relief map of the Ural region.

Scientists of Bashkir State University have been examining a large plate found in 1999, with a picture of the region done according to an unknown technology.

It is a real relief map.

Today's military has almost similar maps.

This ancient map of an unknown civilisation is a Civil Engineering Works Map of South Ural.

The map contains civil engineering works: a system of canals with a length of about 12,000 kilometres, weirs, and gigantic dams. Not far from the canals, diamond-shaped grounds are shown, whose purpose is not known. The map also contains numerous inscriptions. The inscriptions were executed in a hieroglyphic-syllabic language of unknown origin. Scientists have not been able to read it.

The find was made by a doctor of physical and mathematical science, Professor Alexandr Chuvyrov of Bashkir State University.

Since the 18th century, a number of reports had been filed concerning 200 unusual white stone slabs covered with signs and patterns. These were situated not far from Chandar village in the Nurimanov region. It was these reports that set Chuvyrov on his search.

The first slab was found on July 21, 1999 under the porch of a house in the village. It was 148 cm high, 106 cm wide and 16 cm thick (59 inches high, 42 ½ inches wide and 6 ½ inches thick). It weighed at least a ton.

Says Chuvyrov: "At first sight, I understood that this was not a simple stone piece, but a real map. And not a simple map, but a three-dimensional. You can see it yourself."

The slab was studied by a group of Russian and Chinese specialists in cartography, physics, mathematics, geology, chemistry and ancient language. It was found to portray an accurate map of the Ural region, including the Belya, Ufimka and Sutolka

Rivers, as well as Ufa Canyon. The map is on a scale of 1:1.1 km.

The slab consists of three levels. Firstly, the base: this is 14 cm (5 ½ inches) thick, made of the firmest dolomite. The second level is probably the most interesting, "made" of disposed glass. The technology of its treatment is not known to modern science. Actually, the three dimensional picture is marked on this second level. The third level is 2 mm (one twelfth of an inch) thick and made of calcium porcelain. This porcelain covers the slab, protecting the map from external impact.

"It should be noticed," observed Chuvyrov, "that the relief has not been manually made by an ancient stone-cutter. It is simply impossible. It is obvious that the stone was MACHINED." X-ray photographs confirmed that the slab was of artificial construction and was put together with some precision tools.

The longer the slab was studied, the more mysteries appeared. On the map, a giant irrigation system could be seen. In addition to the rivers, there are two 500-metre-wide canal systems, as well as 12 dams. The dams are each 300 to 500 metres wide, approximately 10 kilometres long and 3 kilometres deep. The dams most likely helped in turning water in either side. To create them, over one quadrillion cubic metres of earth was shifted.

In comparison with that irrigation system, the modern Volga-Don canal looks like a scratch on today's relief maps. Dr. Chuvyrov supposes that mankind today can build only a small part of what is pictured on the map.

According to the map, initially Belaya River had an artificial river bed.

What could be the purpose of the map? An investigation was undertaken at the Centre of Historical Cartography in Wisconsin, U.S.A. The Americans were amazed. They concluded that such a three-dimensional map could have but one purpose—a navigational one. They stated that this could be worked out only by means of an aerial survey.

Currently, in the United States, work is being carried out on creating a global three-dimensional map like that. However, the work is not expected to be completed before 2010.

The question is that in the course of compiling such a three-dimensional map, it is necessary to work over too many figures. "Try to map even just a mountain!" exclaims Chuvyrov. "The technology of compiling such maps demands super-power computers and aerospace survey from the Shuttle."

It appears as though those who lived and built at that time used only air transport: there are no roads on the map. Or they very likely used water ways. One opinion is that the authors of the ancient map did not live there at all, but only prepared that place for settlement through draining the land. This is perhaps the most plausible theory.

Further investigations of the map continue to bring one sensation after another. Now the scientists are sure that this map is only a fragment of a bigger map of the earth. According to some hypotheses, there were a total of 348 fragments like the one being studied. The others are thought to be possibly in the region where this one was found.

In the outskirts of Chandar, the scientists took over 400 samples of soil, which led them to conclude that the whole map had been most likely situated in the gorge of Sokolinaya Mountain (Falcon Mountain)—but that during the glacial epoch it had been torn to pieces. It is thought that if the scientists manage to gather the "mosaic," the map should have an approximate size of 340 by 340 meters (1,105 by 1,105 feet).

After having studied archive files, Chuvyrov ascertained some approximate places where four pieces could be situated: one could lie under one house in Chandar, another under the house of merchant Khasanov, the third under one of the village baths, and the fourth under a pier of the local narrow-gauge railway bridge.

(*Pravda* news release, April 30, 2002)

ASTRONOMY

3. Knowledge of planet beyond Pluto

Ancient Sumerian scientists held that there was one more planet beyond Pluto. (An old map showed the sun and eleven planets, counting the moon, all by size.)

There is now good reason to believe that this "secret" planet does exist. The outer planets Uranus and Neptune have been seen to wriggle slightly in their orbits around the sun. Pluto has been eliminated as an influence. Scientists now speculate that the culprit may be an undiscovered planet's gravitational pull.

On December 31, 1983, scientists announced their belief that they had located a body in orbit beyond Pluto.

4. Calendars more accurate than ours

Ancient Maya employed calendars more accurate than ours: The Maya calculated 365.2420 days to a year; our Gregorian calendar calculates 365.2425 days to a year; according to present-day astronomy, the actual figure for the duration of the year is 365.2422 days to a year.

In other words, the Mayan calculation was closer to the sidereal figure than is our present calendar. These ancient Central American Indians had a more precise calendar than we in this "age of science"! The Mayan year was accurate to nearly 1/10,000 of a day. Their surviving calendars are absolute proof that they fixed the duration of the day correctly to three places of decimals.

METALLURGY

5. Bronze harder than we can make

Bronze is a hard alloy made of copper with the addition of 1/10 part tin. (By distinction, brass is an alloy of copper and zinc.)

The early Chinese, as well as the Canaanites (inhabitants of Canaan, the land now known as Israel) knew how to harden bronze to the strength of high-grade steel (harder than we can produce)—we still do not understand this technique.

6. Iron that will not corrode

In Egypt, instruments made of non-corrosive iron ("arms which did not rust") which had been buried in vaults were reported by the Arab historian Ibn Abd Hokm.

In China, iron farm implements were discovered in the tomb of Emperor Ch'in Shih Huang in Sian Province, that were not rusty after 2,000 years in wet soil.

7. Enormous castings of pure iron (we can't)

At Mehauli, near Delhi, India, ancient castings of LARGE pieces: The Ashoka Pillar (1,500 years old) is a column of cast iron 6 tons and 23 feet 8 inches high—a huge casting job, with *hardly a trace of rust.*

Here is testimony to a sophisticated unknown science. (Iron that was 1 ,500 years under tropical heat and monsoon should have corroded and disappeared long ago.) This is pure iron, which *can be produced today only in tiny quantities and by electrolysis.*

At Kottenforst, in West Germany, is an iron pillar much older, likewise weathered, but with very little trace of rust.

8. Alloy processes not yet discovered

The Mochicas of Peru produced alloys of gold, silver and copper—worked by processes not yet discovered.

9. Copper hardened by an undiscovered process

On display in an Ecuador museum is an ancient, steel-hard copper wheel.

Numerous artifacts recovered from ancient mounds in Michigan, U.S.A were likewise made from chilled, or hardened copper by a method long lost to mankind.

In Egypt, copper-headed chisels have been discovered that were tempered in some manner unknown today.

10. Silver that does not tarnish

In Ecuador, ancient utensils have survived, fashioned in silver that does not tarnish, even to this day.

11. Gold plating technique superior to ours

From Costa Rica to Peru, a gold plating technique was used that required fewer operations than our present-day method.

These are some particulars in which ancient metal-lurgy remains in advance of our own. It seems certain that metallurgy declined and became forgotten; we are still attempting to rediscover its secrets.

CONSTRUCTION: SIZE AND TECHNIQUES

12. Architecture beyond our scope today

Examples of such construction ability are numerous. We need only visit Tiahuanaco, in Bolivia, South America, to see evidence of this. Things that can't happen have happened here. The site is built 12,000 feet above sea level. This is oxygen-poor air, in which the slightest exertion can cause nausea and worse. Yet blocks of up to 200 tons were manoeuvered over distances up to 90 miles. In rarified air this is not possible by muscular strength. This grand complex was built with a technical skill embarrassing to us and by a method unknown to us.

Here was a city of startling dimensions.

Acres and acres lie covered with truncated pyramids, artificial hills, lines of monoliths, platforms, underground rooms and giant gates which incorporate architecture beyond our technical scope at the present day:

Many large gateways were built from a single stone. The Gate of the Sun is the biggest carved monolith in the world, a single block 10 feet high and 6 feet wide.

The size of some of the buildings is astounding. The "Sun Temple" was on a platform 440 feet long by 390 feet wide, composed of blocks 100 to 200 tons each. Walls of the temple complex itself had blocks 60 tons each. Steps of the stone stairway were 50 tons apiece. The paved court of the Palace is 80 feet square. Court and hall are one single block of dressed stone.

• Building blocks: one is 36 feet long and 7 feet wide, fitted without lime or mortar and without any joint showing.

13. Lifting capabilities beyond ours—*engineering feats never equalled since*

In Peru, the ruined mountaintop "fortress" of Sacsayhuaman (pronounced "sexy woman") overlooks the ancient capital of Cuzco. Its terrace walls are 1,500 feet long and 54 feet wide.

Within a few hundred yards of the complex, an abandoned *single block the size of a 5-story house* weighs an estimated 20,000 tons! Yes, 20,000 tons. It is impeccably cut and dressed. We have no combination of machinery today that could dislodge such a weight, let alone move it any distance. This indicates mastery of a technology which we have as yet not attained.

The quarries are 20 miles away, *on the other side of a mountain range and a deep river gorge.* How the gigantic stones were moved across such hopeless terrain is anyone's guess.

Whilst in Peru, we could visit Ollantaytambo. Here there are ancient fortress walls. This walls are composed of tightly fitted blocks weighing 150 to 250 tons each. The blocks are of very hard andesite. Special tools are required to penetrate such *hard* rock.

The quarry is on a mountaintop 7 miles away. At a 10,000-feet altitude, would you believe, the builders carved and dressed the hard stone, lowered the 200-ton blocks down the mountainside, crossed a river canyon with 1,000-foot sheer rock walls, and then raised the blocks up another mountainside to fit them in place.

Away on the other side of the world is Lebanon. The site of ancient Baalbek conceals a mystery that may never be solved. Two magnificent Roman temples were built upon an already existing, immense, prehistoric dressed platform. These temples, the greatest in the Roman world, were dwarfed by the platform. The platform is a feat of engineering that has never been equalled in history.

• Here are individual stones as big as a bus. Up to 82 feet long and 15 feet high and thick, they are estimated to weigh 1,200 to 1,500 tons each. One block weighs 2,000 tons—4 million pounds of solid rock! It contains enough material to build a house 60 feet square and 40 feet high with walls a foot thick.

• And you notice that they are raised into the building as much as 20 feet above ground.

34

- There are tunnels in the walling large enough for a train to go through.

Even with the tools of modern technology, we could not move these building blocks intact. Our largest railway cars are too puny. There are no cranes or other lifting apparatus in the world today that can budge, let alone lift, these titanic blocks—yet they are fitted together with such precision that no knife blade can be inserted between the blocks.

It would take three of our largest overhead cranes (hoisting 400 tons each) to lift one of them—even if it could be done without damaging the block by the stress of its own tremendous weight. At freight-train speed, the largest freight car can transport just 110 tons. Supposing that somehow a block could be maneuvered onto a wheeled vehicle, the enormous load would drive the wheels into the ground or grind them to pieces on the rock surface.

One individual block still lying prepared in the quarry is 12 feet high by 12 feet thick and over 60 feet long. To move it by brute force to join the others would have taken the combined efforts of 40,000 men. (But then how could so many have had access to the slab, in order to raise it?)

Now I shall take you to one of my favorite spots on earth. Perched in the Andes mountains of South America, on a razorback high above a horseshoe canyon, Machu Picchu, in Peru, is a breathtakingly beautiful site. These fabled ruins instill romance and mystery.

But just notice those squared blocks—they're 16 feet long! And look above these doors—each granite lintel weighs 3 tons.

Now to Easter Island, in the Pacific Ocean. Here we are on a most isolated island.

Hundreds of mysterious stone faces, each weighing 35 to 50 tons, jut from the soil and stare out to sea. They once wore red hats. The hats alone weighed 10 tons apiece, had a circumference of 25 feet, a height of 7 feet 2 inches—and were put on *after* the statues were erected.

The statues were carved near the crater top, and then lowered 300 feet over the heads of other statues. This was accomplished without leaving as much as a mark. Then they were moved up and down cliff walls and on for 5 miles to their present resting place.

On a dangerously windy sheer rock face plunging 1,000 feet straight into the sea, is a ledge—400 feet down. On this precarious ledge, 25-ton statues were lowered to stand.

The question is, how did the builders cut, move and erect the gigantic heads, including those which approach *the size of a 7-story building?*

Thor Heyerdahl, the Norwegian explorer of "Kon Tiki" fame, attempted to duplicate the accomplishment of the builders, by brute power, using the "heave-ho" method. With a dozen natives laboring (with increasing frustration) for 18 days, he succeeded in setting up one stone head, and then, satisfied, abandoned the job.

There are several inescapable problems concerning this experiment:

1. The stone head chosen for removal was not a typical-sized stone. (At ten to fifteen tons, as against the others weighing thirty-five to fifty tons, it certainly was a great achievement, but not typical.)

2. His team shifted it a few hundred feet, across smooth sandy ground (which exists only in that location), whereas the other stone heads had to be moved five miles across volcanic rock, hard and uneven.

Heyerdahl's "heave-ho" method, if utilized across such a surface, would have grooved the stones with long scars. None of the original statues show such markings.

3. Heyerdahl's team utilized ropes and wooden poles. However, originally there was no wood on Easter Island. The nearest forest was 2,500 miles away. And ropes made from the local reeds were neither durable nor strong—quite inadequate for such a method. (Fortunately for Heyerdahl's experiment, he used strong, manufactured ropes from Europe.

4. Although Thor Heyerdahl succeeded in moving one small head a short distance over a relatively flat surface, this does not explain how other large heads were moved 300 feet up and down cliff walls.

5. The statues once wore ten-ton hats, put on after the statues were erected. How was this done? If earthenware ramps had been used to erect the hats, they would have had to be several hundred feet long and traces of them would have been found on the island, but no such traces have been found.

14. Larger than our biggest modern buildings

Consider the pyramids at Giza, in Egypt. The largest structure is the Cheops pyramid.

The Cheops pyramid is 476 feet high, with a base 764 feet, and covers thirteen acres (an area almost equal to seven city blocks). The polished limestone facings (now removed) covered 22 acres.

It is still larger than any modern building. New York's Empire State Building is among the very highest erected by modern man, yet it is only about 2/5 the volume of the Cheops pyramid

The Cheops pyramid in Egypt comprises 2,300,000 blocks, totalling 6,250,000 tons in weight (each stone is 2½ tons). This amounts to more stone than has been used in all of England's churches, cathedrals and chapels built since the time of Christ.

Covering the "King's Chamber" are granite slabs of 60 to 70 tons each, brought from a quarry 600 miles away.

The casing stones (which are still in place on the north face near the base) each weigh 15 tons.

Now to eastern Asia. The tallest building in the world until this century (and still the most massive structure on earth) is an ancient pyramid 120 stories high! To see this, we would need to go to China, to the north-western province of Shensi. It can be found on a long, desolate, flat stretch of land about 40 miles west of the ancient capital Sian-fu, on an old dirt-road caravan trail that crosses from Peking to the Mediterranean. *About 2,000 feet at the base, it rises some 1,200 feet high.*

In *Dead Men's Secrets*, I stated that there are actually seven pyramids, flat-topped, with three carved giants resting along the outer edges. Further discovery has now revealed that there are more than 20 pyramids in that area.

The four faces of the pyramids are, like so many ancient structures, aligned to the compass points.

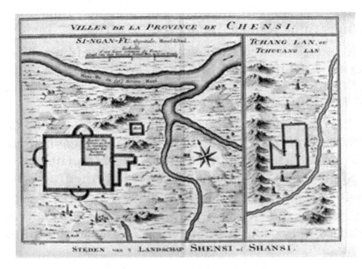

Traces of color remain on the sides, indicating the colors that were given to each side: east—aqua green; south— red; west—black; north—white; and on the flat tops—traces of yellow.

A pair of American adventurers who roamed Asia between the two world wars, R. C. Anderson and Frank Shearer, were shown these pyramids. (Anderson visited Egypt's pyramids in 1970 and believed himself to be the only man living to have seen both the Chinese and the Egyptian pyramids.)

In 1946, a U.S. Army airplane crew rediscovered and photographed these pyramids from the air.

Place 26 skyscrapers the size of the Empire State Building and you have the volume of the largest Shensi pyramid.

A report in Nexus, October-November 1995, referred to two other travellers in 1912, Fred Meyer Schroder and Oscar Maman, who gathered information about these same pyramids, "mountains as high as the sky". The largest would have a volume 20 times as large as the Great Pyramid at Giza, Egypt.

"The sides are now partially covered with trees and shrubbage," they reported. "It almost looks natural hill. We rode around the pyramid, but did not discover any stairways or doors."

A US Airforce map of the area, compiled using satellite photographs, shows no less than 16 pyramids. This part of China (anciently called Sian-Fu, but now known as Xian) was the heart of the ancient Chinese empire.

In October 1994, Hartwig Hausdorf climbed one of these pyramids and was able to count 20 more pyramids in the immediate vicinity. This really flabbergasted him, because no one knew of their existence.

Says Hausdorf: "China has still a lot of mysteries—even the local population quite often isn't aware of them. It's a small miracle I received the go-ahead to enter some 'no go' areas. I was, in fact, the only one who was granted such favours. I assume there are two reasons for this. I regularly visit China with a group of tourists. In 1993, I became acquainted with Chen Jianli, an avid researcher of his country's past. He assured me he would try and open a few doors inside the Chinese Ministry of Tourism. In fact, in March 1994 I was able to visit some former 'no go' areas in the Shaanxi-province. I passed around some copies of my German book, Die Weisse Pyramide (The White Pyramid) to the right people. I talked to archaeologists who at first denied any pyramids, but finally recognised they did exist."

It was that month (March, 1994) that Hausdorf met with Professor Feng Haozhang, Xie Duan Yu and three colleagues. They denied the existence of the pyramids, until Hausdorf showed them three photos of three different pyramids. At this, they "caved in".

The following October, they gave Hausdorf permission to enter some other 'no go' zones.

So at last, after many decades of rumor, photographic evidence was now available. Says Hausdorf: "Most scientists denied the existence of pyramids in China. If any scientist still clings to that, show him my photographs."

15. Surface smoothing and fitting accuracy superior to ours

We return to Sacsayhuaman, in Peru. Here enormous blocks (up to 25 feet wide and of 50 to 200 tons) are so intricately flush one to the other, it is impossible to pass a knife blade between them.

One block in an outer wall has *faces cut to fit perfectly with twelve other blocks.* There are other blocks cut with as many as ten, twelve and even *thirty-six sides*—and with no mortar between them. Each fits exactly to the next touching stones, from every side, *including the inner surfaces!* It defies belief. The whole system interlocks and dovetails, making the chance fitting of each block, or the grinding back and forth in situ for a perfect fit impossible. Even if it had been possible, the power required to do this would be sufficient to supply the needs of a modern city. Do you see the problem?

Back again to Macchu Picchu, high in the Andes mountains. We enter a room. Each wall is composed of a single solid megalith, carved into thirty-two angles which join it to the neighboring blocks perfectly. Such walls astound modern architects.

In Cuzco, Peru, we see the same phenomenon. . . walls composed of multi-shaped blocks with every side dovetailing perfectly into adjoining blocks. I photographed one with 14 visible sides, each of which interlocks exactly with adjoining stones.

Consider again the Cheops pyramid, which is perfectly square to within 3/10,000 percent.

Although it is constructed of 2,300,000 great blocks put together without any cement, you still can't get the thinnest blade of a knife between them. The joints of the original limestone casings are "barely perceptible, not wider than the thickness of silver paper." So observed archaeologist Howard-Vyse, who uncovered part of the original limestone casings near the base of the pyramid. One of today's biggest U.S. contractors has stated that we do not possess any machine capable of making equally smooth surfaces as those connecting the stones of the pyramids. They were fitted to an accuracy of 1/100 inch.

The pyramid is *level* over an area of 13 acres to within half an inch.

It is the world's most accurately aligned building, true north.

16. More accurate standard of measurement

The Cheops pyramid incorporates higher mathematics in its very design, and advanced scientific knowledge in its measurements. The relationship of the pyramid's height to the perimeter of its base is the same as that between the radius and circumference of a circle. It thus incorporates the mathematical value known as *pi* (the constant by which the diameter of a circle may be multiplied to calculate its circumference)—and it does so accurately to several decimal places. Its main chamber made use of several

Pythagorean functions not "discovered" supposedly until thousands of years later.

It served also as a calendar by which the length of the year can be measured to the exact minute. And it was as an observatory from which maps of the stellar hemisphere could be accurately drawn.

It is so finely aligned to the North Pole that modern compasses can be adjusted to it.

> The measurements of its sides and angles accurately reflect the geographic measurements of the northern hemisphere, such as the degree of latitude and longitude, the circumference and radius of the earth—even accounting for polar flattening. All this data was not 'discovered' until the seventeenth century. (Jeffrey Goodman, *Psychic Archaeology*, p.97)

Ken Johnson, in his book *The Ancient Magic of the Pyramids* (page 89) amplifies this point:

45

"Expressed another way, it allowed them to translate the four curved quadrants of 90 degrees that form a hemisphere onto the flat surfaces of four triangles.

> The result is that the pyramid's height represents half the earth's polar diameter, and its perimeter represents the equatorial circumference. In the same proportion, the pyramid's total surface area represents that of the northern hemisphere. For all intents and purposes, the Egyptians 'squared the circle' and 'cubed the sphere'...
>
> It is now clear, for example, that the Great Pyramid was built with a base length representing the distance that the earth rotates in half-a-second.

Herodotus wrote that the exact slant height of the pyramid is one stadium long, that is, one sixhundredth of a degree of latitude.

Agatharchides (second century B.C.) reported that the length of a side was one-eighth of a minute of a degree of latitude.

(In the International Geophysical Year in 1958, the exact dimensions of the earth were determined by satellite, and the French meter—which is our own standard system of measurement, supposedly based on the dimensions of the earth—was found to be incorrect. But more amazingly, the Egyptian cubit—the unit of measurement used in the pyramid—was found to be exact. In other words, the cubit fits into the dimensions of the earth within five decimal places—a rather startling coincidence.)

17. Construction speed superior to ours

The Great Pyramid was *erected at an incredible speed.* Recent evidence suggests that this enormous structure may have been built in a fraction of the time generally assumed. It may have been built in 4 years by just 4,000 workers, laboring only 3 months a year!

An inscription in the Pyramid of Snofru (Sneferu), which is two-thirds the volume of the Great Pyramid, shows it took only two years to construct. By similar methods, the Great Pyramid would have been completed in as little as four years.

Furthermore, excavations at the Great Pyramid have uncovered only 4,000 laborers' huts, which in no way could have housed 100,000 workmen.

The problem that emerges for conventional historians is how only 4,000 men could build the Great Pyramid with 2,300,000 blocks in just four years, during just three months each year, if wooden sledges and barges were used.

This is a technological feat beyond comparison in the modern world. The supposition that enormous manpower, inclined planes and rollers were used, must be discarded.

If we are to believe in the use of wooden sledges and barges, mathematicians tell us that 26 million trees would be required just to fashion the necessary number of sledges and rafts—more than Lebanon or the ancient world could have supplied in the twenty years we are told the job took. Two tomb paintings of the Twelfth Dynasty which show sledges and barges being employed to transport a few statues were concerned with methods used, not in the Fourth

Dynasty (when the Great Pyramid was built), but a thousand years later in the Twelfth Dynasty.

Hieroglyphics from the different dynasties indicate a decided decline in the technology and life-style of Egypt after the time of the Great Pyramid. This is supported also by the funerary texts in the Book of the Dead. The Egypt of the history books, with which we are familiar, was but a vague shadow of the supergreat early Egypt.

Something else. To handle or move one of the blocks might require a thousand hands (500 men), for whom there would not have been room around the stone. (Assuming the use of primitive methods, the block must still be handled, even if only to pass ropes under it, or to load it onto a barge.)

Furthermore, engineers have estimated that a ramp to service the Cheops pyramid would finally have had to be a mile long, with a volume of masonry four times greater than the pyramid itself. No, that's not how they built it.

18. Buildings which are virtually indestructible

In a search for hidden powers and riches, Melik al Aziz, in 1196, employed thousands of workers to pull down the three Giza pyramids stone by stone. They went at the smallest pyramid for 8 exhausting months, after which he gave the order to suspend all work when he saw that the building had scarcely been touched.

The pyramids are as strong today as when they were built. Scientists have conceded that modern man cannot build a great pyramid that would retain its shape for thousands of years without sagging under its own weight.

I ask you, what kind of people were they who knew so much more than we do today of engineering, and who constructed giant edifices that still stand? These are construction miracles that have not been repeated.

The twin towers of the World Trade Centre in New York, prior to the terrorist attack on September 11, 2001, were said to be indestructible. It took just two aircraft to bring them down. Without any doubt, the Great Pyramid is very much more solid. Humanly speaking, it IS virtually indestructible.

19. Shaking towers—a secret unknown

Have you heard of the Shaking Towers of India? In Ahmedabad, Gujerat, two minarets, 70 feet tall and 25 feet apart, have a peculiarity that is unique in the world. If a small group of people sets one tower in motion by a rhythmical to-and-fro movement, the other tower begins to swing too. (Secret unknown. The roots of the science behind this are buried deep in time.)

20. Cement superior to modern Portland cement

Now here is a secret that was known all over the world, it now appears—from remains discovered in Yucatan, Mexico; California, U.S.A.; from the seabed off Bimini; from Ecuador, Malta, Peru and Egypt.

The remains are of a crystalline-white, flint-like building glue, nearly identical, yet superior, to "modern" Portland cement (superior in its unique combination of two qualities: fast-setting speed and exceeding strength).

Recent examination of many prehistoric buildings has led to detection of this glue. Traces of iron oxide rust where the glue grips the stonework suggest that iron oxide was added to the cement. Iron oxide grows fingers or hairs at high speed to form a fast-setting, tenacious interlocking network.

In Egypt, the stone blocks of the Great Pyramid were joined together with cement that was PAINTED ON like paint.

Modern chemists can analyse this cement, but are unable to reproduce it.

Yet this cement, as thick as a sheet of tissue paper, can hold a vertical joint about 5 feet by 7 feet in area.

The cement is so tough that the limestone blocks will break before the cement gives way. So fine is the join that a photograph fails to reveal it.

21. A process for softening hard rock

There is evidence that in ancient times, both in Peru and Tahiti, a softening procedure for hard rock was in use, enabling it to receive hand or foot imprints by pressure only, as though the granite was putty soft.

How did they do it? Quite possibly from a plant extract! It seems that this *process for softening hard rock, by utilizing a radioactive plant extract,* may have been used by the Incas and others in shaping stones.

• An earthenware jug discovered in a Peruvian grave contained a black viscous fluid that, when spilled on rocks, turned them into a soft, malleable putty.

• American archaeologist A. Hyatt Verrill saw remnants of this substance in the possession of an Indian witch doctor.

• Fawcett. the British explorer, recorded in his diary that on a walk along the River Perene, in Peru, a pair of large, Mexican-type spurs was corroded to stumps in one day by the juice from a patch of low plants. The plants grew about a foot high and had dark-reddish fleshy leaves. A local rancher commented: "That is the stuff the Incas used for shaping stones."

• A small, kingfisher-like bird in the Bolivian Andes bores holes in solid rock by rubbing a leaf on the rock until it is soft and can be pecked away. This bird is probably the white-capped dipper (cinclus leucocephalus), which nests in spherical holes, on the banks of mountain streams.

It appears that ancient races discovered and used this fascinating secret—something that modern science has not yet learned to apply.

Gary Webb of South Weston, New South Wales, Australia, wrote to me recently:

> While working with two Greek bricklayers in the Blue Mountains, they told me a story of someone from their family.
>
> The story goes that they were helping to excavate an archaeological site when they came across a type of glass container with liquid inside.

They opened it up and poured the liquid out onto a rock, which caused a hole to go straight through the rock.

This happened somewhere in Greece (near Athens, I think, but I'm not sure).

22. Non-circular, 5-point star shaped holes drilled

From the seabed at Bimini, ancient stone building blocks have been salvaged and brought to Fort Lauderdale, Florida to be used in the construction of a jetty.

One intriguing feature of these blocks is that some of them had been already drilled through. But get this. The drill holes show a feature that is most unusual.

Here we see perfectly drilled holes with sharply defined tips cut right through 12-foot long blocks of granite—while the hole, right through the rock, is not round (as would result from a modern drill), but shaped like a 5-pointed star! Also other huge 1½ to 6 inch diameter round holes perfectly drilled through 12 foot thick blocks. (Richard Wingate, *Lost Outpost of Atlantis*, pp. 154, 162–165)

(Granite, one of the most abrasive stones, wears expensive diamond drill bits down to nothing very quickly.)

23. Drills faster than modern power drills

Around the turn of the twentieth century the great Scottish Egyptologist Sir William Flinders Petrie was honest enough to admit that he was puzzled by certain

CONSTRUCTION: SIZE AND TECHNIQUES

granite and diorite 'drill cores' that he had found at ancient Egyptian archaeological sites. It was clear that tubular drills had been used to hollow out large blocks of these peculiarly hard stones and that the drills had been turning *with great force and speed.*

The amount of pressure, shown by the rapidity with which the drills pierced through the hard stone, is very surprising. A load of at least a ton or two was placed on the 4-inch drills cutting in granite. On the granite core No. 7, the spiral of the cut sinks one inch in the circumference of 6 inches, or one in 60, a rate of ploughing out which is astonishing.

These rapid spiral grooves cannot be ascribed to anything but the descent of the drill into the granite under ENORMOUS PRESSURE.

Just how enormous that pressure may have been was recently confirmed by Chris Dunn, an expert machine toolmaker, who now occupies a senior position at Danville Metal Stamp in the United States. His calculations indicate that the ancient Egyptian drills were turning "500 times faster than modern power drills."

So, in addition to the reliefs and texts of ancient Egypt, we have here hard empirical evidence of the form of the drill hole cores, which suggests that the ancient Egyptians may have had access to some sort of as yet unidentified power source which they harnessed in a variety of ways and for a variety of purposes.

24. Ability to slice through hardest materials without friction or heat

Talmudic-Midrashic traditions speak of a most unusual building tool, called Shamir, that was used in the construction of the Temple of Solomon in Jerusalem. This device was capable of cutting the toughest materials "without friction or heat", was reportedly "noiseless", and could slice through diamonds.

Special precautions surrounded its use: "The Shamir may not be put in an iron vessel for safekeeping, nor in any metal vessel; it would burst such a receptacle asunder. It is to be kept wrapped up in a woollen cloth and this in turn is to be placed in a lead basket filled with barley bran."

The Shamir was best known by its function as "the stone that cuts rocks", although it was often also referred to as a "worm" or a "serpent".

Judaic tradition directly associates the Shamir with Moses. (Graham Hancock, in *Daily Mail*)

25. Interlocking tunnels thousands of miles in length under land and sea

This is the most astonishing and most suppressed archaeological secret: the existence of inexplicable tunnel systems beneath the surface of a great part of the earth. These are part natural and part artificial.

Stories of mysterious subway systems exist in the legends, folklore and myths of almost every country. Reports have persisted for thousands of years.

There is good reason to believe that a gigantic system of *interlocking tunnels thousands of miles in length* extends under Ecuador and Peru. It also connects Lima to Cuzco, and goes on to Bolivia, or the sea. Many hundreds of miles have been explored and measured. Ingeniously constructed entrances are masked beyond discovery; there are *elaborate devices* to trap robbers and hidden doors of carved stones with no sign of a crack or joint. The tunnels are so imposing that some conjecture them to be the work of an unknown race of giants. The Incas, at the time of the Spanish threat, deposited much of their treasure in these caves and sealed some of the entrances.

Accounts of these tunnels come initially from the Spanish invasion of 1531. Searching for the entrances was recommenced by the Peruvian authorities in 1844, after a dying Quiche Indian (a direct descendant of the Incas) confessed the secret of the tunnels to a priest.

Theodore Roosevelt, later to become U.S. president, picked up accounts of these sophisticated prehistoric tunnels during his expedition in 1914.

The famous British missionary-explorer David Livingstone reported on subterranean excavations 30 miles long in Rua, North Africa.

In the ruins of Tiahuanaco in Bolivia, the nineteenth century naturalist Charles d'Orbigny also saw the entrances of galleries leading to a secret underground city.

Natives of Ecuador and Colombia speak of tunnels with *cut-stone walls as smooth as glass* in the mountains.

In August 1976, Scotsman Stanley Hall led a seventy-strong team to investigate one section of the Ecuadorian tunnel system. The expedition was supported by the Universities of Edinburgh and Quito, with assistance from the British and

Ecuadorian armies, and accompanied by no less a celebrity than the astronaut Neil Armstrong.

The party fought its way up the raging torrent of the Rio Santiago to arrive at the shaft where, 700 feet below, the entrance to the tunnels lay.

Similar strange tunnels, very ancient and of unknown origin, were brought to the attention of Christopher Columbus in Martinique, in 1493.

Indications of the reality of these tunnels come also from Sweden, Czechoslovakia, the Balearic Islands and Malta. Most ancient tunnel entrances are now covered by landslides.

These were the constructions of men with an advanced knowledge of engineering.

26. Canals and dams larger than ours

For information on these, see Item no. 2.

27. Construction methods beyond our present capability

There are reports, claims and allegedly ancient maps of prehistoric tunnels that not only extend beneath the earth, but even run under the oceans – for example, under the sea from Spain to Morocco, and under the Atlantic and Pacific Oceans (See *Dead Men's Secrets*, ch.19, items 22, 37, 47, 50, 64, 68, 69 and p. 205)

Of all responses to my book *Dead Men's Secrets*, the chapter on ancient tunnels has attracted the most interest. Naturally enough, the mention of large systems of tunnels has elicited skepticism from some readers. How could the ancient races have had such technology? Why haven't we heard more about these tunnels? With all our advanced technology, even we ourselves have not accomplished such things! And so on. Some have accused me of telling wild tales.

Indeed, for some time, I had to agree on one point: that most of these fantastic tunnels were constructed in ways beyond our present capabilities—probably by some kind of thermal drill or electron rays, which melted the rock but left no debris.

Beyond our present capabilities? I now have to turn this into a question...because it seems very likely that we are catching up fast—and we may have even caught up—to this particular accomplishment of our ancestors.

Here is the truth: the technology for constructing tunnels hundreds, even thousands, of feet below the ocean floor does now exist. The experience, the expertise, the machinery and the trained personnel are available. And—for secret government projects—even the money is no drawback.

It should be borne in mind that the petroleum industry routinely bores into the deep rock beneath the floor of the ocean. There is evidence available that existing tunnel boring machines are capable of advancing even through fractured rock at an average rate of five miles per year. Given better conditions, it is well within the state of the art to make advances through rock of ten or more miles per year.

With only one machine and crew, a tunnel system 100 miles long could definitely be constructed within a period of 10 to 20 years. Employing five machines, 500 or more miles could certainly be excavated within the same time period.

This capability is real. For example, in the early 1980s a tunnel boring machine was used by the Pacific Gas and Electric Company to bore a 24 foot 1 inch diameter, 22,000 foot long tunnel. This was for construction of the Kerckhoff 2 Underground Hydro-electric Power plant, about 30 miles north-east of Fresno, California. Typical rates of progress were from 60 to 100 feet per day. If one assumes 365 days of work per year, this would achieve about 5½ miles of tunnel per year. (Edward R. Kennedy, P.E., "The Kerckhoff 2 Underground Hydroelectric Power Plant Project, A State-of-the-Art Application of a Tunnel Boring Machine", US National Committee on Tunneling Technology, *Tunneling Technology Newsletter*, number 38, June 1982)

At a meeting with government leaders of South Korea, the then Japanese prime minister Yoshiro Mori proposed a 108 mile tunnel under the sea to link the two countries. Mr Mori stated: "The construction is technically possible, but the problem is money." ("Japan proposes undersea tunnel to link S. Korea", http://www.indiatimes.com/221000toi/22worl15.htm, 2000)

Richard Sauder, writing in Nexus, says, "The money to carry out a secret project of this sort certainly exists in the Pentagon's 'black budget'. The requisite infrastructure of secrecy to carry out such a project has been in place in the military-industrial complex for decades now. And there is even a paper trail that shows US Navy interest in building manned bases deep below the ocean floor." (*Nexus*, "The Evidence for Secret Underwater Bases", August-September 2001, p. 28)

All over the world there are mine tunnels that extend offshore under the sea. (Shan-tung Lu,

CONSTRUCTION: SIZE AND TECHNIQUES

"Undersea Coal Mining", paper presented to the
Department of Mining, College of Mineral Industries,
The Pennsylvania State University, University Park,
Pennsylvania, USA, March 1959; George E. Sleight,
"A Hydrographic Survey and Undersea Borings in
Ayr Bay", *Transactions of the Institution of Mining
Engineers*, vol. 112, 1952–1953, pp.521–541; J.T.
Robertson, "Drifting Under the Firth of Forth",
Canadian Mining Journal, December 1964, pp.70–71)

A United states Navy document from 1966 forth-
rightly discusses major military installations
constructed beneath the sea bed. It states that

> Large undersea installations with a shirt-sleeve
> environment have existed under the continental
> shelves for many decades. The technology now
> exists, using off-the-shelf petroleum, mining,
> submarine, and nuclear equipment, to establish
> permanent manned installations within the sea
> floor that do not have any air umbilical or other
> connection with the land or water surface, yet
> maintain a normal one-atmosphere environment
> within....
>
> ...a Rock-Site installation consists of a room or
> series of rooms, excavated within the bedrock
> beneath the sea floor, using the *in situ* bedrock as
> the construction material. (Richard Sauter, *Under-
> water and Underground Bases*, chs. 5 and 6. Adven-
> tures Unlimited Press, USA, 2001)

It should be noted that these "rooms" under the sea
are probably enormous. (An underground power
plant in the Himalayan Mountains of Bhutan is
hundreds of feet long and over 100 feet high! Lloyd A.
Duscha, former Deputy Director of Engineering and
Construction for the US Army Corps of Engineers in
Washington, DC, said in a public speech:

There are other projects of similar scope, which I cannot identify, but which included multiple chambers up to 50 feet wide and 100 feet high using the same excavation procedures...

Sauder notes,

Tunnelling under oceans, seas, bays and estuaries has been done for a very, very long time, all over the world, stretching way back at least into the 19th century, if not before. Undersea tunnels can stretch for miles and reach depths of 2,000 feet or more beneath the ocean floor. Of course, today's technology is far more powerful and sophisticated than it was 50, 100 or 150 years ago. One can only speculate as to how long, how deep and how elaborate contemporary, clandestine, submarine tunnels might be. (Ibid., p.29)

And perhaps there exists an even more sophisticated tunneling technology than what we are being told about. After 30 years researching high technology, Norio Haykawa reports:

"In 1950, we had some amazing tunneling machines, using nuclear devices that could go 10 miles an hour melting the hard, tougher rocks and creating tunnels." He refers to military tunnels allegedly linking facilities in Nevada with others in California. ("The Prophecy Club", *Secrets of Dreamland*—video on top secret government research centers)

Tunnels extending for long distances under land and sea? Let's get real. It happened in the past and it is happening again!!!

TOWN PLANNING

28. Town planning for centuries ahead—nowhere done today

The Toltecs of Mexico undertook building projects covering 400-years-ahead planning. (At present no country in the world has any planning for centuries into the future.)

MECHANICAL-ELECTRONIC

29. Robot cock, long buried, able to shriek and flap its wings

An automaton in the form of a cock was discovered after the Moslem conquest of Egypt, by one of the first caliphs of Cairo, and described in an Arab manuscript known as the Murtadi. It was made of red gold and covered with precious stones, with two shiny gems for eyes. The cock, when approached, uttered a frightening cry and began flapping its wings.

EVERYDAY ITEMS

30. Books composed of gold leaves

Ancient books made of gold leaves are known to have existed in Guatemala and Egypt.

CLOTHING, TEXTILES

31. Cotton grown in various colours

From Peru to Mexico, cotton was grown in various colors (from brown to blue), a technique which modern science has been unable to reproduce.

32. Textiles finer than is possible in a modern loom

The Incas produced magnificent textiles, with veils, brocades, and 'gobelins'—far finer than is possible on any modern loom.

ART, SCULPTURE

33. Three-dimensional effect from luminous paints

At Ajanta, near Bombay, India, sixth century cave murals portraying women carrying gifts lack depth until the light is switched off. In darkness, the figures on the wall appear to be three-dimensional, as if they were made of marble—by the clever employment of luminous paint, the secret of which has been lost forever.

34. Finest stone engraving in the world

The engraved bones of Glozel in France are still the finest in the world.

35. Mysteriously high carvings

Near Havea, in Brazil, is a *mountain carved to resemble the head of a bearded man* wearing a spiked helmet: On one side of the mountain (on a small vertical face 3,000 feet in height) is a carved inscription in *cuneiform characters* some 10 feet tall. It is a mystery how this was done.

36. Art visible only in a photograph (also visible/invisible art)

Faces that change sex—from "pre-glacial" Germany: In November, 1967, some giant stones resembling human heads were dug up in Hamburg, by an excavator under the direction of engineer Hans Elieschlager. Professor Walther Matthes, a German archaeologist, examined them. Some figures had two faces, and when they were turned 125 degrees, *a man's face turned into a woman's face.*

(Reported in the Russian review *Technique and Youth*, November 6, 1965)

On the Plateau of Marcahuasi, Peru, we find *four-dimensional* art:

Here are found outdoor carvings which, according to the angle of vision, have *several faces*— but you have to move into the right spot to distinguish each of them. As you *move*, they *disappear* or *change* into other figures.

Many become visible, then invisible again, seen only at noon, or twilight, or certain other hours, or at one of the solstices—and *at no other time.*

Visible, then invisible ancient sculpture has also been found elsewhere in South America, as well as in England (at both Stonehenge and Avebury) and in France (southern Brittany).

Among other sculptures on the Marcuahasi Plateau of Peru is a *figure of an old man, when photographed, changes into the carved face of a radiant youth.* How can we explain this sculptural mystery, which is revealed only in the photograph? It is hard to see how an artist could achieve this effect even with the benefits of modern science.

The artists needed incomparable skill to make the shapes appear only from certain angles and under certain specific conditions of sunlight.

These examples belong to a vast complex of monuments and sculptures covering a square mile, using whole cliffs; with images of the four main human races and of animals from other parts of the world.

Every type of sculptural technique was used: bas-relief, engraving, and play of light and shade. The sculptors scientifically utilized *the laws of perspective and optics.*

37. Neon bright "movie" in granite

And now to the Onega River in Russia. Russian archaeologists working in Karelia have come upon what has been termed "prehistoric cinema".

Here is an astonishing work of art: carvings designed to produce a 15 minute moving picture sequence with the brilliance of a neon sign. This amazing gallery of 600 sketches has been incised into a cliff of hard granite rising vertically from lake waters—ingeniously designed to blend the effect of rippling water reflection with setting sunlight.

When the sun nears the horizon, the granite shines dark red and the various colored lines of the pictures become very clear (the countless tiny crystal prisms of the incisions reflecting much more light than the surrounding smooth areas) and shine intensely.

The luminous pictures then begin to move. The frog seems to turn into an elk, while the hunter makes a movement with the hand. Having thrown the axe with his right hand, he puts out his left arm to keep his balance, as the camp fire flickers.

This magnificent spectacle lasts a quarter hour, until the setting sun makes the designs grow weaker.

Like neon lamps going on and off, which seem to move, the same effect is seen here: groups of tiny prisms on the unequal surfaces of the designs act like lamps, so that at moments some become more luminous than others. Two currents of light—from the setting sun and the moving water—give the impression that the whole design is moving. (Konstantin Lauskin, in the review *Snanje-Sila*, cited by Peter Kolosimo, *Not of This World*, pp.204–205)

The artists must have had a clear idea of what they wanted to show, as well as acute vision and steady hands, since a wrong cut in the sharp silica could have ruined the picture for good; *granite is a canvas which will not allow corrections.*

Such subtle techniques run counter to the common view of primitives living in caves, using clubs and crude flint tools and looking like ape men!

So far I have not been fortunate enough to hear an explanation of such ultra modern genius that is even tolerably convincing. Some of this art is so advanced, its techniques are ahead of our time.

38. Crystal skull beyond our ability to produce

I know of at least *two separate crystal skulls*, intricate and beautiful, which have no equal anywhere. One of these was found in British Honduras, the other in Mexico. Each was carved from a solid block of crystal. The larger is in natural size, with a separate, but attached, moveable lower jaw, implying that it was meant to "talk." It is believed to be of Aztec origin. The other is probably of Mixtec origin. The use of these crystal skulls is unknown. Each crystal skull has a built-in, nearly perfect set of complicated optical systems that allow almost all the light entering its base to emanate in full spectrum through the hollow eyes. It has been reported that one skull emits strange noises at times.

One is in the Museum of Mankind (part of the British Museum, London). The other is privately owned.

Today we know that the properties of quartz are very important in semiconductor electronics and radio communications, but with all our experience we could not produce such a work as the crystal skull.

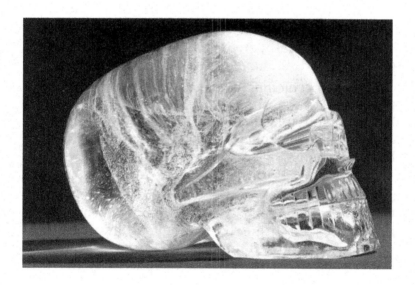

HEALTH, MEDICAL

39. Currently incurable diseases cured with medicines today unknown

Many of today's "incurable" diseases were, in ancient India, actually cured with medicines that to us are unknown. At present, research is being undertaken in Indian texts, in the hope of recovering some of this lost science.

40. Obsidian instruments 1,000 times sharper than today's conventional surgical blades

Sharp surgical instruments made of obsidian and dated to 500 B.C., have been discovered in Peru.

Obsidian (volcanic glass), if you didn't know, is a thousand times sharper than the platinum-plus blades used in other surgical instruments. The cutting surface of obsidian is so sharp that it does not bruise the cells.

In October, 1975, an American doctor performed successful major surgery with obsidian tools. Don Crabtree of Kimberly, Idaho, one of the world's top authorities on lithic technology, allowed a local surgeon to remove a tumor from his lung—entailing the cutting of an incision three-fourths of the way around his trunk.

Obsidian surfaces may eventually revolutionize surgery and could be of special value in cosmetic and plastic surgery.

41. Skull operations technically superior to our own

We now go to Lake Sevan in Soviet Armenia, where thousands of years ago some splinters were removed from the brain of a woman. This lady had received a head blow, in which a blunt object of 1-inch diameter punctured the skull and splintered the inner layers of cranial bone. Surgeons carefully cut a larger hole around the puncture to remove the splinters that had penetrated the brain. Evidence shows that she survived surgery for 15 years.

The discovery was made at Ishtikunny, near Lake Sevan. The examinations were performed by Professor Andronik Jagharian, anthropologist, and director of operative surgery at the Erivan Medical Institute in Soviet Armenia. (William Dick and Henry Gris, "Delicate Head Surgery Was Performed 3500 Years Ago", *National Enquirer*, September 10, 1972, p. 30)

Even by modern standards, such operations would be considered extremely difficult. Some of these operations are seen to be *technically superior* to modern-day surgery.

ELECTRICITY

42. Perpetual lamp, able to burn for 2,000 years

When the sepulchre of Pallis was opened in Rome, in 1401, the tomb was found to be illuminated by a perpetual lamp which had been alight for more than 2,000 years. (Nothing could put it out, until it was broken to pieces by desecrators.)

Kedrenus, the eleventh century Byzantium historian, records having seen a perpetual lamp at Edessa, Syria, which had been burning for 500 years.

We now cross to the Via Appia, Rome. A sealed mausoleum (with a sarcophagus containing the body of a shapely and beautiful patrician girl), opened in April 1485, contained at her feet a lighted lamp which had been burning for 1,500 years! The body was that of Cicero's daughter, Tullia. She lay in an unknown, transparent, allpreserving fluid. When the preserver was removed, her lifelike form with red lips and dark hair was seen by 20,000 people. The lamp continued to burn for some time.

Such lamps functioned indefinitely without oil or any product that was burned or consumed. Touching them was prohibited under pain of provoking an explosion capable of destroying an entire town.

There is no doubt that the ancients knew of energies other than electric power, if they could construct perpetual lamps that burned for hundreds of years. Probably they had more sources of light than we can imagine. Did they utilize chemical power or some form of radiology?

Here's another thought: Lighting by means of the use of hydrogen can be accomplished by a phenomenon called condoluminescence, a cold process. A phosphor is spread on the inside of a tube, similar to the conventional fluorescent lamp. Upon coming in contact with the phosphor, small amounts of hydrogen combine with the oxygen in the air to excite bright luminescence in the phosphor.

That's an interesting fact, but we have to admit that the perpetual lamps of the ancient races are still a mystery.

43. More known about matter, light, and ether properties than we know

It seems that our ancient forbears knew more about matter and light, about the ether and its properties, than the scientists of the twenty-first century can ever know or imagine.

We know how to generate electricity from coal, from water power and from nuclear reactors. These scientists went much further. They were very likely able to extract free energy from the atmosphere for lighting, heating, moving great weights and for instruments used in household chores—not to mention defense.

I am not unmindful of the inventor Tesla. He showed that free energy once used by those ancient civilizations, was indeed possible for us to enjoy today—if we would let it happen. Of course, that would require permission from our gracious benefactors who supply power to us by other means.

44. Paired TV screen transmission

All descriptions of scientific development in China in the first millennium B.C. refer to "magic mirrors". They are mirrors which have extremely complicated high reliefs on the back of the looking-glass. When direct sunlight falls on the mirror, the high reliefs which are separated from the surface by a reflecting glass, become visible. This does not happen in artificial light. If they are set up in pairs, they transmit images, like television. The phenomenon is scientifically inexplicable, by present knowledge.

Some of these mirrors are still supposed to exist in private collections. We do not understand how they were made or what they were used for.

45. Book TV with vanishing pictures

Early this century, Maxim Gorky, the celebrated Russian writer, met a Hindu yogi in the Caucasus, who asked Gorky if he wanted to see something in his album. Gorky said he wished to see pictures of India. The Hindu put the album on the writer's knees and asked him to turn the pages. These polished copper sheets depicted beautiful cities, temples and landscapes of India, which Gorky thoroughly enjoyed.

When he finished looking at the pictures, Gorky returned the album to the Hindu. The yogi blew on it and smilingly said: "Now will you have another look?" Gorky recalls: "I opened the album and found nothing but blank copper plates without a trace of any pictures! Remarkable people, these Hindus!" (Andrew Tomas, *We Are Not the First*, pp. 166–167)

SURVEYING

46. Cross-country lines laid out straighter than by the best of modern survey techniques

On Peru's Nazca plain, pictures traced on the ground are *so large that they can be seen only from the air.* One is 825 feet long! An amazing series of geometric patterns, interspersed with artistically sophisticated pictures of flowers, birds, animals, insects and people, covers 30 square miles.

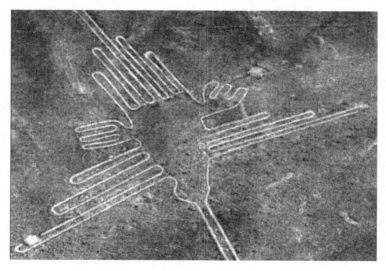

Why were they designed so big? Why were they made in such a way that they are perfectly visible from above, but impossible to spot from the ground? (Thus they were not recognized for what they were until 1939, when a plane flew over them.) Even the original artists could only have recognized the perfection of their creations from the air.

What is more, a deviation of just a few inches would spoil the proportions, which, as we see them on aerial photographs, are perfect.

This is not on a human scale. It suggests, rather, a civilization of titans. And they evidently possessed highly developed instruments for reckoning. There are lines which go for miles, remaining absolutely straight while jumping over (or through) a mountain. They are laid out straighter than by the best of modern survey techniques.

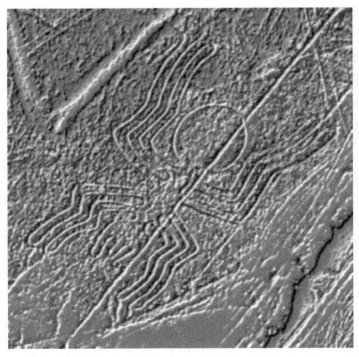

FLIGHT

47. Aircraft able to fly straight up, down, forwards, or backwards

An old Indian document, the *Mahabharata* (written 500 B.C., but referring to a period 1,000 to 2,000 years earlier) makes repeated references to great god-kings riding about in *vimanas* or "celestial cars", described as *"aerial chariots with sides of iron clad with wings"*. They were used both for peaceful transportation and in war.

There is mention of some flying cars which have crashed and are out of action, others standing on the ground, others already in the air.

The vimana was *shaped like a sphere* and borne along at great speed on a mighty wind generated by mercury. It moved in any way the pilot might desire, up or down, forward or backward.

48. Maneuvers that helicopters can perform today only partially

Another Indian work, the *Ramayana* (likewise very ancient) describes one vimana as *"furnished with window compartments and excellent seats."*

"Bhima flew along in his car, resplendent as the sun and loud as thunder. . . The flying chariot shone like a flame in the night sky of summer."

The vimana was a double-deck circular aircraft with portholes and a dome. It flew with the "speed of the wind" and gave forth a "melodious sound". A pilot had to be well trained; otherwise no vimana was placed in his hands. The craft performed maneuvers which only helicopters can do partially today, i.e., stop and remain motionless in the sky.

Detailed descriptions of the ocean and landscape from a great aerial height are given.

The vimanas were kept in hangars, and employed for warfare, travel or sport.

"A vimana rose vertically into the air with a whole family on board, and with a tremendous noise."

49. Ability to hear conversations in enemy planes

The ancient *Samarangana Sutradhara* (also from India) relates that *"the chariot was automatic; big and wellpainted, it had two floors and many rooms and windows."*

This document deals with take-off, *cruising for thousands of miles*, normal and forced landings, and even with possible collisions of aircraft with birds.

The advantages and disadvantages of *different types of aircraft* are discussed at length, as to their relative capabilities of ascent, cruising speed and descent, and recommendations given regarding suitable metals for construction.

Also dealt with are informative details on how to take pictures of the inside of *enemy* planes, methods of determining their approach pattern, means of rendering their pilots unconscious, and how to destroy enemy planes. The secret of making planes invisible, and the secret of hearing conversations and other sounds in enemy planes is documented.

When I first published *Dead Men's Secrets*, and referred to this ancient document, it seemed to me almost incredible that the ancient world could have possessed such surveillance ability. How technology has advanced in the brief span of time since! In truth, if we only knew it, ever since the Soviet satellite Sputnik was launched in 1957, a massive effort has been poured into satellite technology.

As SECW Newsletter (January–February 1995) noted:

SIXTY-FOUR SECRETS STILL AHEAD OF US

There are at least 23 satellites overhead at this moment that have resolution capable of reading the print on a postage stamp lying on the ground. This technology is able to track people, listen to their conversations and watch their movements.

I quote from a report in *Update International*, February to April 1997 (p.16):

Satellites now orbiting the earth are linked to equipment which already can monitor every telephone call you make or receive. Intelligence agencies monitor every local and long distance phone call worldwide...every one of them!

They do it with supercomputers listening for 400 key words. If you say "White House", "bomb", "Clinton", those kind of key words, then they down load it and it is listened to by a human. If you don't use those words, no one ever listens.

In the United States, 'the Justice Department maw is today swallowing a staggering profusion of data about the American people: the location and lifestyles of approximately 80 million households (note: that should be about all of them with a population of some 250 million); transactional information about all toll calls made from hundreds of thousands, perhaps millions, of telephones; substantive information drawn from the secretly recorded conversations of at least 632,000 individuals in just one recent year.' (David Burnham, *Above the Law*. N.Y.: Scribner, 1996, p.168)

"In 1990, 'computers in the FBI inventory could handle 203 MIPS (million instructions per second)'. (Ibid., p.134)

How many times when you replace the phone on the hook do you realise that the actual telephone is still on when you hang up? There are devices which

80

intelligence agencies use, by which they dial a code. Your phone doesn't ring, but everything in your office or home can be heard and recorded.

Even television sets are being equipped with monitors that scan a room every two seconds to record the movements and actions of every person in the room.

Originally designed to test viewing audiences for the benefit of advertisers, the devices could be used to monitor the whereabouts and actions of certain people for surveillance purposes.

An article in *Nexus* magazine, titled "No Place to Hide From State-of-the-Art Satellites" (August–September 2001 issue), citing *Pravda*, July 14, 2001, brings us up to date:

Unknown to most of the world, satellites can perform astonishing and often menacing feats...

A spy satellite can monitor a person's every movement, even when the 'target' is indoors or deep in the interior of a building or travelling rapidly down a highway in a car, in any kind of weather (cloudy, rainy, stormy).

"There is no place to hide on the face of the earth. It takes just three satellites to blanket the world with detection capacity. Besides tracking a person's every action and relaying the data to a computer screen on earth, the amazing powers of satellites include being able to read a person's mind, monitor conversations, manipulate electronic instruments and physically assault someone with a laser beam."

The terrorist attack on the United States on September 11, 2001, sent the world into shock.

It was several weeks before a planned response against world terrorism was initiated with air strikes over Afghanistan.

On October 11, 2001 (precisely one month after the terrorist strikes), in a front page news story, the New Zealand Herald reported:

> "Spy planes equipped with listening devices and superoptic cameras are ready to begin a round-the-clock hunt for Osama bin Laden.
>
> "After quickly gaining mastery of the skies against the ill-equipped Taleban forces, the United States and Britain will draw on the world's most advanced airborne cameras and spying equipment for the next phase of their operation.
>
> "The US has requested use of an RAF Canberra bomber, which has been converted into a spy plane. Its electrooptical camera can take photographs from 48,000 ft. One source said it could identify 'paper in your back garden'.
>
> "Other devices can peer into secret mountain spots from more than 160 km away.
>
> "The aerial surveillance will lead to a more dangerous phase of the military campaign, in which ground forces will pursue bin Laden."

Something else. With ground surveillance radar, it is now possible from miles away, to listen to your car...listen to your footsteps. The technology is now available to scan your retina and those of your passengers as you drive along the road, thus identifying you personally at a distance as you approach.

A type of surveillance camera currently in development boasts the equivalent of x-ray vision and can penetrate clothing to 'see' concealed weapons, plastic explosives or drugs. Known as the passive millimetre wave imager, it can also see through walls and detect activity. (*Update International*, February to April 1998, p.16)

We noted that ancient writings describe [a] the ability "of hearing conversations and other sounds in enemy planes; [and b] the secret of receiving photographs of the interior of enemy planes," and so on.

Now I'm not so sure that we have yet achieved these things in actual practice. Have we yet monitored conversations in enemy aircraft; have we photographed the inside of an enemy plane flying toward us? Certainly I would be interested if any data is available, but until then, I suspect that the ancient world may be still ahead of us in these particulars.

50. Flight powered by mercury

The same ancient document referred to in Point 45, the *Samarangana Sutradhara*, offers a description of the fuel power source for ancient flight:

> "Within it must be placed the mercury engine, with its heating apparatus made of iron underneath.

> "In the larger craft, because it is built heavier, four strong containers of mercury must be built into the interior. When these are heated by controlled fire from the iron containers, the vimana possesses thunder power through the mercury. The iron engine must have properly welded joints to be filled with mercury, and when fire is conducted to the upper part, it develops power with the roar of a lion. By means of the energy latent in mercury, the driving whirlwind is set in motion, and the traveller sitting inside the vimana may travel in the air, to such a distance as to look like a pearl in the sky."

The intricate knowledge of aircraft and flying was deliberately controlled by a select few.

Precautions were taken against industrial espionage and unlicensed manufacture.

The mercury was heated by "a special flame capable of being directed." A laser?

This is not the only ancient Indian document to refer to mercury as a fuel for flight. The *Mahabharata* likewise states that the vimana was borne along at great speed on a mighty wind generated by mercury.

Sir Isaac Newton wrote: "Because the way by which mercury may be impregnated, it has been thought fit to be concealed by others that have known it, and therefore may possibly be an inlet to something more noble, not to be communicated without immense danger to the world." (from a speech delivered at Cambridge in July 1946 by the British nuclear physicist Edward Neville da Costa Andrade, citing Newton)

What it is about mercury that could be of "immense danger" to the world we do not know. Yet it seems apparent that the ancients were well aware of the practical application of mercury.

The secret is still beyond our present technology. Nevertheless, at an international space congress in Paris, in 1959, there was talk of producing an "iono-mercurial engine"; and in 1966 the French looked into launching a satellite powered by a "mercury solar furnace".

Some years ago, in Turkestan, Russian excavators discovered in caves what may be *age-old instruments used in aircraft flight*. These are hemispherical objects of glass or porcelain, ending in a cone, each carefully sealed and each containing *a single drop of mercury.*

A description and illustrations of the pots appeared in the Soviet periodical, *The Modern Technologist.*

51. Specified tricks in flight to deceive an enemy *(and formulas that would revolutionise modern aviation)*

G.R. Josyer, director of the International Academy of Sanskrit Research in Mysore, India, has translated into English the 3,000-year-old *Vymanika Shastra,* meaning "the Science of Aeronautics." (Translation, *Aeronautics, A Manuscript from the Prehistoric Past,* was published in book form by Coronation Press, Mysore, in 1973.) It has eight chapters *(6,000 lines),* *with diagrams, on the construction of three types of aircraft.*

Information covers:

•The design of a helicopter-type cargo plane and drawings for double- and triple-decked passenger planes for as many as 500 people.

•Plans for an aircraft that flew in the air, travelled under water, or floated pontoonlike on the water.

•The qualifications and training of pilots.

•The planes were equipped with cameras, radio and a kind of radar, as well as apparatuses that could not catch fire or break.

•Instructions on how to make the aircraft invisible to enemies, how to paralyze other aircraft, how to create the illusion of a star-spangled sky, how to zig-zag in the sky like a serpent, how to see inside an enemy's airplane, how to spy on "all activities going on down below on the ground."

• Also the correct proportions of certain chemicals which will envelop the aircraft and give it the appearance of a cloud.

RUKMA VIMANA

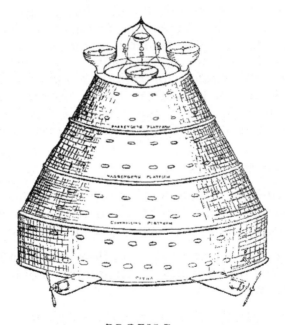

PROFILE

Drawn by
T. K. ELLAPPA,
Bangalore.
2-12-1923.

Prepared under instruction of
Pandit SUBBARAYA SASTRY,
of Anekal, Bangalore.

This document contains formulas which would make our aircraft manufacturers gape in astonishment, and if mastered would herald a new era in aviation.

52. Flight powered by sound waves

Indians of the La Paz region of Bolivia say that thousands of years ago their ancestors travelled in *great golden disks which were kept airborne by means of sound vibrations at a certain pitch, produced by continual hammer blows.*

Similarly, ancient Sanskrit books state plainly that aircraft could be driven solely by the power of sound, tunes and rhythms.

Mention of the relationship between sound waves and antigravity, comes to us also from ancient Babylon, Egypt, India, Finland, from the Bantu of Malawi and the Algonkin of North America.

"Soon after the Spaniards discovered the city [Tiahuanaco, in Bolivia], a Jesuit wrote that 'the great stones one sees at Tiahuanaco were carried through the air to the sound of a trumpet.'" (*The World's Last Mysteries*, p.138)

Tablets indicate that by means of sounds, into the air heavy rocks were raised, which a thousand men could not have lifted. (Francois Lenormant, *Chaldean Magic*)

Certain Arab sources contain curious tales about the manner the manner in which the earlier pyramids were erected. One says that the huge stones were wrapped in papyrus, then struck with a rod. They became weightless and moved through the air. Coptic writings indicate that the blocks were levitated by the sound of chanting.

We detect here traces of an ancient power technology more advanced than our own.

Interestingly, in modern automobile tests, an ultrasonic reactor attached to a carburetor almost doubled petrol mileage with very little exhaust gas. This simple gadget was based on a system of harmonic resonance within the atomic structure of liquids.

Some years ago, NASA scientists were experimenting with moving small particles by a procedure called *acoustic levitation*. They succeeded in using sound waves to levitate pellets of glass or metal. (*OMNI* magazine, November 1980)

53. Electromagnetic energy from the atmosphere

Various ancient writings hint at propulsion achieved by some form of anti-gravity, such as electromagnetic energy, obtained directly from the atmosphere. (See *Dead Men's Secrets*, Chapter 26, item 29; Chapter 27, item 1; and Chapter 28, item 25).

A number of scientists believe that electromagnetic energy supersedes the orthodox laws of gravity. The fundamental characteristics of gravity still elude analysis by modern physics.

Were you aware that magnetic currents can be used not only to neutralize gravity but as motive power?

Spinning objects made of selected materials have been shown to generate an electromagnetic energy field when placed in rapid relative motion. If this force field is made to undulate, a secondary gravitational field is produced, which can "neutralize" earth's gravity.

Disk airfoils three feet in diameter and incorporating an electrical condenser charged with 150 kilovolts have been made to fly under their own power. The disks moved under the influence of interaction between electrical and gravitational fields, in a fifty-foot-diameter course.

An electromagnetic force field would cause the vibratory frequency of an object to be altered in such a way that it opposes the frequency of the gravitational field. It would act upon all parts of the craft simultaneously. A vehicle thus propelled would be able to change direction, accelerate to thousands of miles an hour, or stop.

Do you see, man now uses the sledge-hammer approach to high-altitude, high-speed flight. He has to increase power in the form of brute thrust many times over in order to achieve just twice the speed.

By his present methods, man actually fights against the forces that resist his efforts.

But in using a gravitic field to provide the basic propulsive force, he would make his adversary work for him. If the coupling effect between gravity and electricity can actually be harnessed and used for propulsion purposes, then we would have a free and inexhaustible power supply.

If we could thus conquer gravity, the headaches of transmission of power from the engine to wheels or propellers would cease to exist. The oil-consuming car engine and hydroelectric generators would become obsolete. The work of the world could be done with tiny amounts of energy. Construction of large buildings and bridges would be revolutionized by temporary induced weightlessness.

I think you'll agree, we seem to have lost something.

54. Vehicles adapted to travel both in the sky as well as on or under the sea

The *Vymanika Shastra* from India (the 3,000 year old document referred to under Point 46) contains eight chapters of plans for craft that could travel in the air, on water or *under the sea.*

A similar account comes to us from China. It is reported that more than 2,000 years ago a ship appeared on the sea at night with brilliant lights which were extinguished during the day. It could also sail to the moon and the stars, hence its name, "a ship hanging among the stars" or "the boat to the moon".

This giant ship which could travel in the sky or sail the seas was seen for 12 years.

INTRIGUING SECRETS

Some of the data now to be presented may seem preposterous to the mind educated to a blind faith in modern science. Yet I have reservations about scientific dogmatism, because what is science today will be fallacy tomorrow and what was science ten or twenty years ago is fallacy today.

Certainly, as evidence piles up of a forgotten prehistoric science, it is difficult to shake off the feeling that our ancestors knew a lot more than we do.

They possessed superior intelligence and technological skill—often to a degree that the modern mind finds staggering.

Their attainments stare us in the face; their secrets defy us. Consider these as food for thought.

55. A container which weighs the same whether full or empty

According to writings of early Egypt, a mysterious vase of red crystal, when it was filled with water, *weighed the same as when it was empty.*

56. Fire which burns in water

Fire that burned in water was used by the Greeks to defeat the Arabs in 674 and in 716, and then the Russians in 941 and in 1043. This was a devastating weapon: in the battle of 716, 800 Arab warships were totally destroyed. The secret formula was brought to Greece by a fleeing architect who had excavated it in ancient Baalbek.

This viscous product has never been reproduced, even by modern napalm specialists.

57. Singing statues

• The statue of Memnon, in southern Egypt, as well as its twin, emitted a thin, high-pitched sound like a chord on a harp. It was a phenomenon heard at dawn, for about 200 years; the sound at first was said to be sweetly melodious. (One theory suggests a complex mechanism hidden in the depths of the statue, activated by the rising sun working on a lens hidden in the figure's lips.)

• Similar sounds were heard at sunrise in the granite cave at Syene.

• Also in the Karnak temple.

After certain repairs on the statue of Memnon, the musical sounds ceased. This indicates that the "music" was due to some complicated mechanism triggered by the sun's rays—a mechanism which was inadvertently damaged during the restoration work.

58. Large public baths heated by the flame of a single candle

At Isfahan, Persia, large public baths were heated by a crucible which was in turn heated *by the flame of a single candle!* What was this ingenious mechanism that could amplify the energy of the fire of the candle thousands of times?

59. Objects suspended indefinitely in mid air

During the fourth century B.C., in Greece, a vault was constructed with "magnetic stones" so that idols could be suspended in mid-air.

Around 400 A.D., at Alexandria, Egypt, a sun disk ascended into the air "by magnetism" in the temple of Serapis.

Similarly, in a temple in Hierapolis, Syria, Lucian (155 A.D.) personally witnessed an image of Apollo raised into the air.

He reports: "Apollo left the priests on the floor and was borne aloft."

A century later, it was reported that in Asia Minor an iron Cupid was observed suspended between the ceiling and floor of a Diana temple.

Just a few of numerous other reports include:

93

- In Tibet, an embalmed body poised a span from the ground.

- At Bizan, Abyssinia, a "flying rod" suspended motionless for centuries, mid-air, in a church.

- At Heliopolis, Egypt, a statue was maneuvered into position while floating in the air.

On the basis of the foregoing, should we not reconsider the method by which those enormous stone blocks strewn throughout the world were set in place? The explanation offered by the ancients—who lived closer to the events than we—cannot lightly be brushed aside. Even the legends of their local descendants should at least be given a hearing. Outrageous though individual accounts may sound, a comparison does reveal an interesting thread of agreement running through them all—the concept that the stones were made to levitate.

Egyptian Coptic writings indicate how the blocks were *levitated by the sound of chanting.*

Babylonian tablets assert that sound could lift stones.

The relationship between sound and weightlessness is still a mystery to us. We are reminded, again, of the Bolivian legend that flying vehicles were kept airborne by vibrations at a certain pitch generated by continual hammer blows.

Here, perhaps, is the explanation for those technologically impossible prehistoric constructions, many of which seem literally to have been thrown up to the tops of mountains and perched on the edges of precipices, as if the giant stones had flown there!

However, at our current level of knowledge, we can establish no link between sound and weightlessness.

It is possible that in a particular village in present-day India we see demonstrated the principle behind this ancient secret. Stones are levitated when the right number of people place their index finger on a rock and do the correct chanting. Success in this instance appears to hinge on sound waves and bio-currents from the fingers.

60. Alchemy (transmuting one metal into a different metal)

Reports of alchemy (the transmutation of metals, including lead into gold) come to us from widespread sources: for example, Egypt, the Arabs, China, India, France, England, Switzerland and Sumeria.

The fact of an advanced culture and technology in protohistory can clarify why ancient alchemists believed in transmutation of the elements.

A remote age during which nuclear science was practiced (see *Dead Men's Secrets*, chapter 29) implies the use of atomic energy for many purposes. Some ideas, such as transmutation, which the alchemists kept alive in their endless search to turn lead into gold, most likely stemmed from ancient knowledge that manipulation of atomic structures could convert one element into another.

Today, with instruments such as the strong-focusing synchrotron, the transmutation of metals has begun to look quite good.

It has now been reported that Soviet scientists have found a cheap way of converting lead into gold. They were conducting an experiment in nuclear bombardment when they found the lead shielding inside an advanced nuclear reactor had changed into gold. They were able to repeat the process under laboratory conditions.

Russian scientists have also grown real diamonds in a carbon dioxide bath under low pressure.

But such breakthroughs are *possible only in the most sophisticated contemporary physics.* The idea that such processes were known thousands of years ago, and then forgotten, jolts the mind.

61. Invisibility apparatus

According to writings from ancient Greece, a "magical" helmet, when placed on the head, rendered the wearer invisible. (Was this "helmet" an electronic device to diffract or deflect light rays, thereby acting as a protective agent?)

Among the British Druids, a "magic mist" was created, to render the producers invisible. (This might have been linked with light diffraction devices.)

Ancient manuscripts preserved in Llasa, Tibet, are said to reveal the secret of "antima" (the "cap of invisibility").

In German tradition, the heroic Siegfried won from the dwarf king Alberich a cloak that when worn rendered him invisible. He used it afterward in successful duels. The ancient Germans firmly believed in a cloak of invisibility.

Ancient Indian texts reveal "the secret of making planes invisible."

According to both ancient Indian writings, the *Mahabharata* and the *Ramayana*, weapons and flying objects were able to render themselves "invisible to the enemy."

62. Visible then invisible, bridge

In his book *Timeless Earth*, Peter Kolosimo speaks of a city in the Andes protected by a rocky defile which could only be crossed by a bridge. Constructed of ionized matter, this pre-Incan bridge was made to appear and disappear at will.

Of course, all solid objects have a vibration frequency within the range perceptible to the human eye. Some scientists consider that it might be possible to alter the vibration frequency into vibrations outside the visible range.

In October 1943, secret experiments allegedly took place from the Philadelphia Naval Yard in which an electromagnetic force field was created and a ship of the U.S. Navy, fully crewed, was made to disappear. In other words, ship and crew became totally invisible. Pulsating energy fields produced an electronic camouflage. The explanation can probably be found in Einstein's unified field theory.

Despite detailed evidence as to names, places and times, the event is still officially denied.

Yet, significantly, since then the Great Powers have developed infrared cannons sensitive not only to the visible shape of objects, but also to their thermal radiation...indicating that they seriously consider the possibility of war between invisible combatants.

In particular, the United States has been investigating means of providing an electronic "cloak of invisibility" for its aircraft carriers at sea. The enemy would then attack an electronically depicted but nonexistent target.

A product of similar research is the Stealth aircraft.

63. Time viewing devices

The construction of the "Al Muchefi Mirror" was outlined according to the laws of perspective and under proper astronomical configurations—a mirror in which one could see a panorama of Time. So states researcher Andrew Tomas, in his book *We Are Not the First* (p.167).

If this be true, then the former cultures were one step ahead of us—they had Time Television.

In his book Readable Relativity. the British scientist Clement V. Durell writes: "But all events, past, present and future as we call them, are present in our four-dimensional space-time continuum, a universe without past or present, as static as a pile of films which can be formed into a reel for the cinematograph."

Astronomers tell us that in this wonderful universe the past is still in the present, depending on where one is in space. How can this be? A simple illustration will suffice.

We see a star by the beam of light it sends out. That light beam takes a certain length of time to arrive here. Meanwhile the star moves on through space. By the time that first beam of light reaches us, and we see it, the star has moved to a different part of the sky. But we see that star as though it were still in the same place as when the beam of light left it.

We look, for example, at the Andromeda nebula; the light we see left it 820,000 years ago; we are therefore looking 820,000 years INTO THE PAST! Astronomers behold a star "explosion". We say it has just happened, that we have seen a star "explode". But they know that we are looking at something that happened hundreds or even thousands of years ago, and we are just receiving the rays from it, just as those rays left that star when it was happening.

Recent studies seem to indicate that FROM EVERY POINT IN THE UNIVERSE, WAVES GO ON, RECORDING WHAT HAS HAPPENED. Rays from this earth are endlessly going on out into space, recording what we have done here on this earth.

Over these rays we have no control. They are the result of our actions. We cannot bring them back.

But, if we could travel fast enough, we might perhaps be able to catch up to them and see past events before our eyes.

If we go to Orion, where rays from this earth take 500 to 600 years to arrive, and then look back, perhaps we see Joan of Arc, in 1429, in Orleans, waving her flag.

If we travel to Arcturus and look back, we see President John Kennedy, in 1963, slumping forward from the assassin's bullet.

Perhaps we go to 61 Cygnus, and looking back eleven years we intercept the signals of the Gulf war in Kuwait.

We look back from Sirius and see ourselves, you and I as we were nine years ago, and whatever we were doing at that time.

It's as though invisible signals, like radio waves, were being thrown from us into the heavens. Should we be able to race ahead and catch up with them in outer space, the past would be as in the present; past events would be seen as clearly as though they were still happening to us.

Did you ever wonder what ultimately becomes of the waves that radio and television stations send out into space 24 hours a day? Do they fade and vanish, or do they keep going forever? It is known that on occasion pictures do appear mysteriously, long after a broadcast is finished.

C. B. Colby reported in the White Plains, N.Y., *Reporter Dispatch* an incident that occurred in Britain in September, 1953, and was later mentioned in *Reader's Digest*:

> Suddenly in many parts of England television screens blossomed out with the identification card and call letters of T.V. station KLEE in Houston, Texas. Even today transatlantic programming is but a dream, so several viewers took pictures of the image to prove the happening.
>
> What really startled the T.V. world was the fact that when British broadcasting engineers advised KLEE in Houston of the unusual event, they were told that the station had been off the air since 1950. No KLEE identification card had been shown for the past three years.
>
> Where had that picture been for three years? Why did it appear only in England and how did it get back from wherever it had been?

Did our predecessors have the technology to tune into past events at will? To retrieve vibration waves from those events, which were still rippling somewhere through space?

64. Brain transplants

Brain transplants, the ultimate in neurosurgery, appear to have been carried out in several ancient cultures – namely Peru, India and Sumeria—according to written and pictorial evidence.

IMPOSSIBLE, of course! That's one's instant response to such a suggestion. Good material for a science fiction horror, that's all.

Could it be that we are too hasty to dismiss that which we do not understand? Some recent astonishing developments make skepticism look almost naive.

Dr. Robert White of Cleveland Western Reserve Medical School has transplanted an isolated brain into the body of another monkey in an apparently successful operation.

Russian experimenters have temporarily succeeded with total head transplants on dogs.

The major problem in a transplant of this nature is to make the neuroconnections that will enable the brain to be supported by and to control its new body. Clearly, success in the West is still years away; the difficulties are so immense.

And in a human transplant, powerful moral and sociological questions arise. May I boldly suggest that a brain transplant recipient is himself probably no longer alive? It is the memory, the character, the personality—yes, the very life of the donor that survives. Well?

Now for some startling news. The ultimate breakthrough may have already occurred. In northeast China, during April 1984, a team of micro-surgeons successfully transplanted the head of a corpse onto the body of a living man. The 31-year-old recipient had a massive brain tumor and was being kept alive on life support apparatus. The head was taken from a man who had died after he was almost decapitated in a factory accident in Shensi Province. The surgical team used newly developed, computer-controlled, microlaser techniques.

The fourteen-hour operation was described by one of the team, CAT-scan specialist Chen Lee, who later fled to Europe. From his thirteen notebooks, he planned to write a book about the mind-boggling successes of China's recent transplant experiments.

Then on July 7, 1986, Soviet medical journals reported an experiment at a research facility near Moscow, in which surgeons had switched the heads of two "prime specimens." Although both young men emerged from the surgery with their senses intact, attempts to reattach the spinal cords of their new bodies were unsuccessful and the men were paralyzed from the neck down.

German cancer specialist Dr. Hans Frankl, in Russia to treat victims of the Chernobyl nuclear disaster, was horrified to learn from Soviet doctors that the experiment had been preceded by at least 14 failed attempts.

In Europe, researchers and theologians were quick to condemn the experiments on apparently healthy subjects as "an outrage" and a "godless disregard for human life."

Ancient neurological skills now become more believable, don't they?

We'd love to have you download our catalog of titles we publish at:

www.TEACHServices.com

or write or email us your thoughts, reactions, or criticism about this or any other book we publish at:

TEACH Services, Inc.
254 Donovan Road
Brushton, New York 12916

info@TEACHServices.com

or you may call us at:
518/358-3494